HMS DOLPHIN

GOSPORT'S SUBMARINE BASE

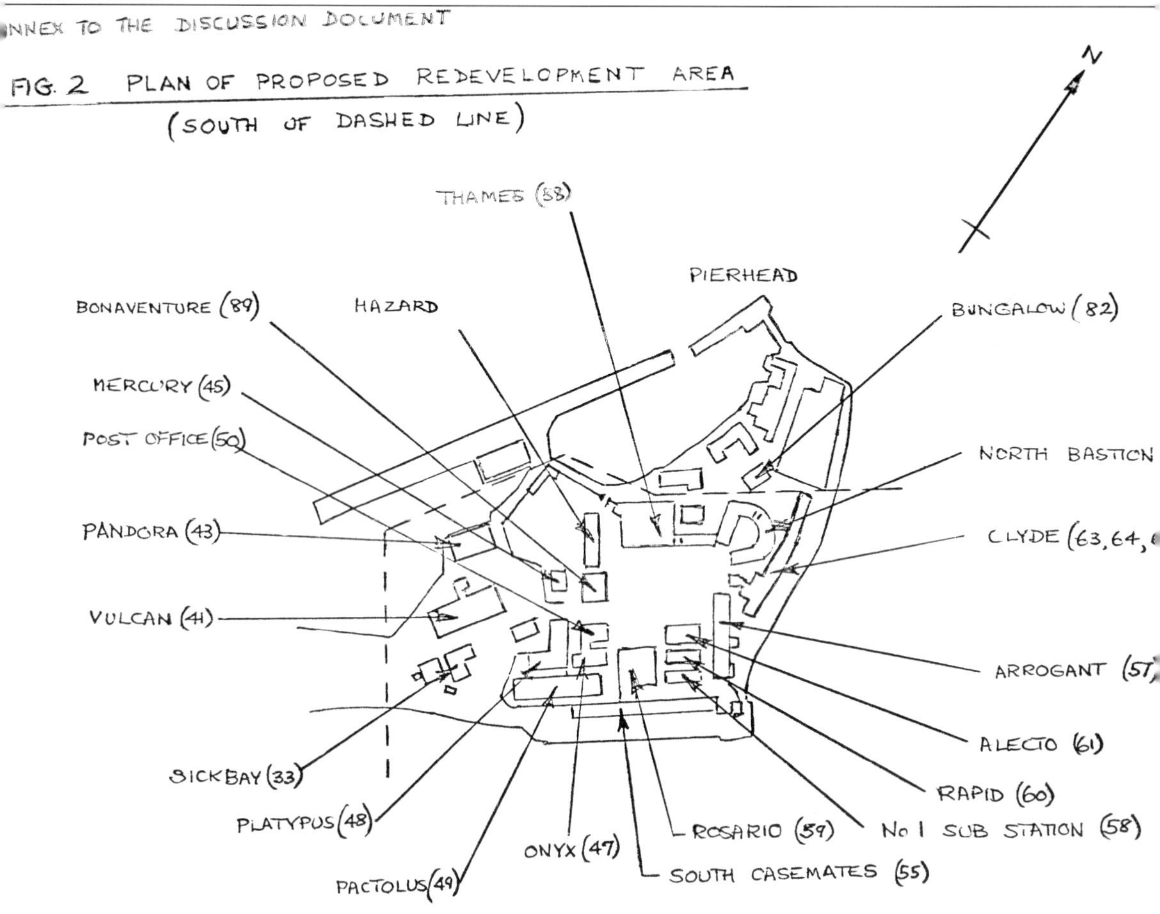

NNEX TO THE DISCUSSION DOCUMENT
FIG. 2 PLAN OF PROPOSED REDEVELOPMENT AREA
(SOUTH OF DASHED LINE)
N
THAMES (58)
PIERHEAD
BONAVENTURE (89)
HAZARD
BUNGALOW (82)
MERCURY (45)
POST OFFICE (50)
NORTH BASTION
PANDORA (43)
CLYDE (63, 64,
VULCAN (41)
ARROGANT (57,
ALECTO (61)
SICKBAY (33)
RAPID (60)
PLATYPUS (48)
ROSARIO (59)
No 1 SUB STATION (58)
ONYX (47)
SOUTH CASEMATES (55)
PACTOLUS (49)

Plan of the Fort area in the 1970s.

HMS Dolphin

Gosport's Submarine Base

Keith Hall

All royalties form the sale of this book are being directed to the
Royal Naval Submarine Museum.

TEMPUS

Tempus Publishing Limited
The Mill, Brimscombe Port,
Stroud, Gloucestershire, GL5 2QG

ISBN 0 7524 2113 1

Typesetting and origination by
Tempus Publishing Limited
Printed in Great Britain by
Midway Colour Print, Wiltshire

ACKNOWLEDGEMENTS

I would like to thank the following for their help and assistance in compiling this book:
Lt Cdr D.N. Parkinson and Mrs I.F. Stephenson from Property Management at Fort
Blockhouse, for not only allowing me to rummage in their photographic archives, but to
make off with the contents. Mr M. Reeves. Cdr Tall and his staff at the Royal Naval
Submarine Museum, Gosport, particularly Mrs Debbie Corner, the very helpful and patient
'Keeper of Photographs'.

Special thanks must go to Mr Brian Woods for the information and pictures of the SETT. I
am also thankful to Mr and Mrs Alan McLelland, Andy Lawrance, Grant Mahoney, Josh
Tetley and Derek Lilleman and John Taylor for the dits, photographs, encouragement and
proof reading assistance.

I am particularly grateful to the organisations below, for permission to use their photographs
and for waiving of their copyright fee:
Crown Copyright, for all the MOD photographs.
RN Submarine Museum for the photographs on pages 20, 21, 22, 26, 27 28/29, 30, 31, 32, 34,
35, 36, 120.
The Imperial War Museum, for the photographs on pages 127.
Gosport Library for the photographs on pages 13, 14, 15, and 18.
British Library for the picture on page 10.
Cdr Tall for permission to use photographs from his book 'Submarines in Camera' written in
conduction with Peter Kemp.
The Navy News for permission to use the photographs on pages 23, 127, 126 which appeared
in 'More Navy in the News 1954 – 1994.

And the Royal Navy, without whom, none of this would have been necessary.

And finally I ask forgiveness of any contributors who may have been inadvertently omitted
from these acknowledgements.

Contents

Acknowledgements   4

Introduction   7

1.   Pre 1904   9

2.   1904 – 1918   19

3.   1918 – 1945   25

4.   1945 – 1960   37

5   1960 – 1998   73

6   Submarine Squadrons   119

An aerial view of Fort Blockhouse, 1918. 'First the Nab, then the Warner, Outer Spit and Blockhouse Corner.' This little poem has been recited by submariners for almost a century, and represents a novel form of navigation. Luckily, the fledgling submarine base was not built at Littlehampton.

The Royal Naval Submarine Museum

The Royal Naval Submarine Museum exists as the living monument to a proud service that has made, from its inception in 1901 to the present day, a significant contribution to the security of the Nation, both in peace and war.

The Submarine Service has been, and remains, an elite and somewhat self-contained world; as a result the role of submarines can be misunderstood, underestimated, or neglected. As the public shop window of the Service, the Museum as the opportunity to correct this through fostering interest and providing facilities for education and research. This places on the museum the responsibility to maintain an institution of the highest possible quality in terms of presentation and organization.

In addition to its feature as a memorial to those who have passed on, it serves as a focus for living submariners, whether retired or still serving. It contains voluminous records on individual submarines and submariners and houses a wealth of diverse artifacts. The Trust provides charitable funds for ex-submariners and their people in need and their Museum thus has strong 'regimental' attachments that must be nurtured and enhanced.

Introduction

It is said that once on a bright and sunny day, at the gates of heaven, St Peter was welcoming the new arrivals from earth, when he noticed one of them causing quite a commotion. The Archangel went across to confront the noisy newcomer.

St Peter: 'Name please Sir?'

Newcomer: 'Nausea'

St Peter: 'Last earthly occupation?'

Newcomer: 'Master at Arms, in Her Majesty's Royal Navy'

St Peter: 'And what appears to be the problem?'

Newcomer: 'For my last two years I was drafted to HMS *Dolphin*, the Navy's Submarine School at Gosport. I decided to move my family with me, and to be honest, that was the last thing I should have done. Those filthy submariners have ruined my life completely. Within the space of a few short months my eldest daughter, who was barely fourteen, was pregnant to a killick stoker off the *Odin*. My son, who was at Holbrook School, and had passed all his exams for Dartmouth, has scrubbed round it all and wants to be a forendie on an 'S' boat. To cap it all, my wife ran off with the Submarine Squadron Chaplain!'

St Peter: 'Dear me, you really have been through it, haven't you?'

Newcomer: 'You ain't kidding, sunshine.'

St Peter: 'And tell me, what caused your demise, Sir?'

Newcomer: 'A broken heart, you moron, what else!'

St Peter: 'Quite, but that doesn't explain what all this fuss is about.'

Newcomer: 'Simple really, if there's any submariners here, I'm off.'

St Peter: 'Submariners here! Outrageous, you must be joking! There's about as much chance of finding a submariner here as a lawyer!'

The Master at Arms eventually accepts St Peter's word and in due course draws his grat issue of harp, wings, the long dress and pillowy cloud. Peace reigns for several months until one afternoon when the very angry and agitated Master At Arms bursts into St Peter's office.

Newcomer: 'You lying, double-dealing bastard!'

St Peter: 'I beg your pardon?'

Newcomer: 'You gave me your word that submariners couldn't get in here and I've just seen one!'

St Peter sinks into his chair deep in thought. After a few moments he begins to chuckle.

St Peter: 'Could you describe this submariner to me?'
Newcomer: 'Of course I can: long, untidy, unwashed hair, with a straggly set. Manky submarine sweater, dirty gown, down-at-heel steamers tied with string, floating on a cloud that stinks of diesel and smoking a roll-up. Whoever he is, he's definitely a bloody submariner. I'm off!'

St Peter, laughing in a very disarming way, puts his arm around the by now very confused Master At Arms.
St Peter: 'It's alright, he's not a submariner, that's Jesus – he just thinks he's a submariner.

While the accuracy of this yarn is, at best, somewhat questionable – after all, who's heard of a Master At Arms with a heart – it does perhaps begin to give an insight into the elitist, sexist and somewhat self-contained world of the submariner, the thinking man's matelot. The submariner, who accepts the hardships, as he accepts the menaces, of his life, generally without complaint. The submariner whose rewards are very rarely measured in purely financial or tangible terms, but to him they are real enough. Such in essence is the submariner – he is brought into the trade with an initiation ceremony that would leave a mason speechless, which probably explains why you can always tell a submariner, you just can't tell him much.

As with other unfathomable and mystical cults, the Submarine Service, along with its own codes and rituals, had its own sacred retreat. Until recently this role was filled, both spiritually and physically, by HMS *Dolphin*, Fort Blockhouse.

I hope this book in some way may help you remember the Navy as it was, not what it became.

Keith Hall
Tumbledown

Bibliography

A Short History of Fort Blockhouse – Cdr W E Warner D.S.C. RN
Fort Blockhouse – Richard Compton-Hall
Portsmouth Harbour Defences
Submarines in Camera – Tall/Kemp

One
Pre-1904

King Edward III first authorized the fortification of Portsmouth Harbour in 1342. Work did not start until 1431, probably due to the shortage of government funds. It was at this time that the first fortifications were built on the Gosport side of the Harbour, on the spit of land where Fort Blockhouse now stands. The harbour needed to be protected on both sides of the inlet, although the fortifications on the Gosport side were on a more modest scale. In the late 17th century the first defences of old Gosport were built, and these were strengthen in the late 18th century as part of a general improvement of the area's defences. During the 18th century Haslar Hospital and the Priddys Hard Magazine were built and the brewery was redeveloped into the Royal Clarence Yard.

A blockhouse, housing five guns, was first build on the Gosport side of Portsmouth harbour in 1495. Henry VIII ordered that it be replaced with an eight-gun battery as part of his Device Forts, in 1539. A treaty between France and Spain in 1538, after Henry's divorce from Catherine of Aragon, left England politically isolated. Fears of invasion were rife, and considering the Kings usual divorce settlements this may not have been an over reaction. Although the threat was short-lived, it caused the largest defence programme since Saxon times. Henry took a personal interest in the military engineering techniques of the time, and approved and amended the designs himself.

It is highly likely the original fort had disappeared by 1667 when Bernard de Gomme installed a twenty-one-gun battery for Charles II. This fortification was not an enclosed fort but a simple battery of about ten guns facing the south, and two or three facing south-west. The structure was repaired in 1668 and 1679. The fort fired a salute to Charles II as he sailed into Portsmouth harbour on 5 September 1683 to inspect the fortifications. Unfortunately, one of the guns exploded and killed one of the crew.

In the early 1700s de Gomme's successor, Sir Martin Beckman, proposed to build new strengthened walls facing south east toward Haslar. Little seems to have been done, despite the War of the Spanish Succession. Disparaging reports by Admiral. Byng and George Talbot, about the condition of the battery at Blockhouse, spurred the government into action, and by 1710 work was well under way with 70 men employed in building the fort. On completion the fort's main gate was set in the south west rampart and was approached by a draw bridge over a moat. At the south end, the moat turned right between the original de Gomme battery and the sea. A wooden palisade formed the remaining two sides of the fort. Little work was done to the fort until the late 1700s, when (probably prompted by the Revolutionary War with France) plans were submitted to reinforce the fort.

The main gate was rebuilt during 1813, and by 1825 the seaward battery had been covered converting it into a casement battery. The present stone north bastion replaced the northern corner of the fort during the early 1840s. An interesting report from 1858 reads: 'The water supply is derived partly from well, partly from rainwater. The privy is dark and noisome at the time of our inspection. It is only flushed at spring tides. It would be a great improvement to convert this privy into a water latrine and to flush it out daily'. Some things don't change!

In 1845 plans were proposed that would bring major improvements to the fort. The parapets were thickened and living accommodation was improved. Although progress seems to have been slow the fort was essentially in its final state by 1863.

In August 1859 the Royal Commission was instructed to consider the defence of Portsmouth. The report was submitted in 1860, and the recommendations were:

1) To prevent landing form the enemy on the Isle of Wight.
2) Protection of the anchorage at Spithead.
3) Defence of the Needles passage.
4) Protection of the harbour mouth.
5) The land defences – these were divided between the Gosport defences and the hill forts.

Fort Blockhouse was updated. Five new batteries were constructed along the sea front and there were six new forts built: Gomer, Grange, Rowner, Brockhurst, Elson and Fareham.

The fort became the home of the Royal Engineers' Submarine Mining School in 1873. The Engineers were responsible for electric light, which, if the mines were electronically activated, would explain their involvement. Being essentially a defensive application, similar 'Mining Schools' were set up in other ports (Sheerness, Plymouth, Queensferry, and Milford Haven). At Fort Blockhouse the pier, that was later to become known as the Petrol Jetty, was built and buildings were erected on the land around where the tank is now sited.

In 1904 the site, considered obsolete by the Royal Commission, was turned over to the Navy.

A sketch by De Gomme shows his battery at Blockhouse Point.

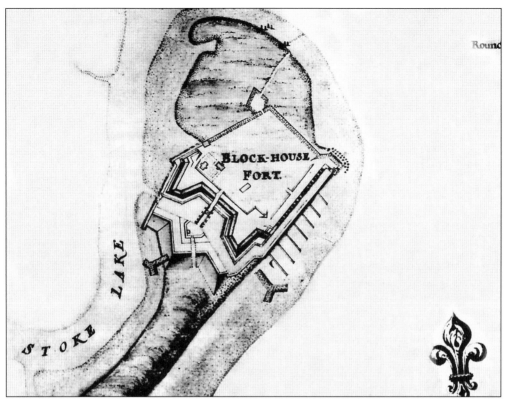

Fort Blockhouse mid eighteenth century.

Fort Blockhouse and Gilkicker Tower 1729.

The west demi-bastion of Fort Blockhouse in the 1830s. Haslar hospital can be seen in the background. In earlier years criminals were hung on the point in front of the bastion; an early and innovative example of a crime prevention scheme.

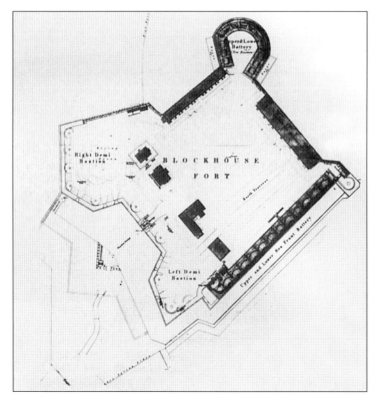

The fort in the early 1860s.

Gosport main gate 1841.

Haslar Gate 1896.

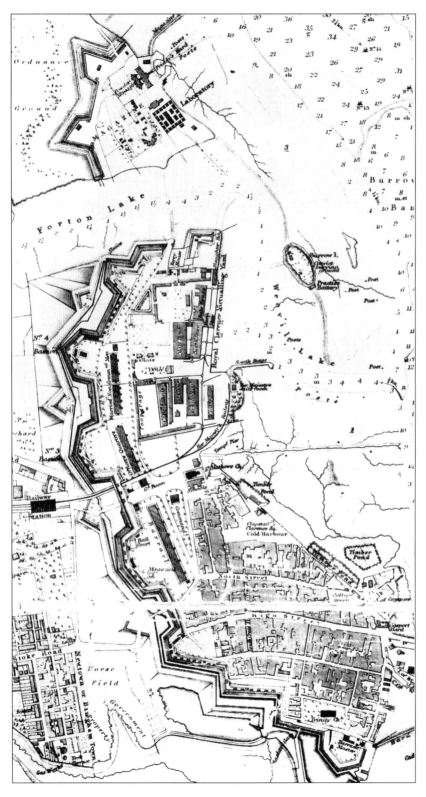

Gosport Lines
1870s.

Haslar Gate from the Gosport side.

Haslar Gate looking towards Gosport.

The Royal Engineers conducting a submarine mining exercise off Fort Monkton 1879.

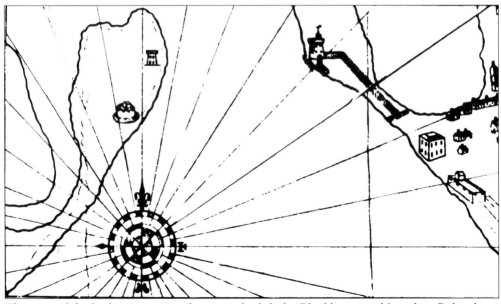

This map of the harbour entrance shows, to the left the Blockhouse and Lymdens Bulwark.

During 1930 several links of the harbour boom chain were excavated on the beach to the east of the Round Tower Portsmouth.

The harbour boom being tested in the early 1900s.

DRINKERS' GUIDE TO OLD GOSPORT

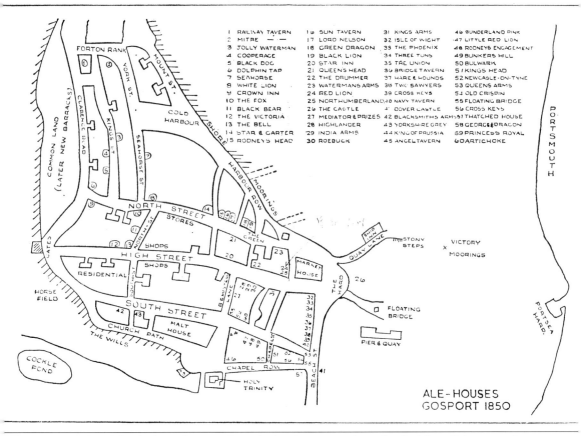

1 RAILWAY TAVERN	16 SUN TAVERN	31 KINGS ARMS	46 SUNDERLAND PINK	
2 MITRE — · —	17 LORD NELSON	32 ISLE OF WIGHT	47 LITTLE RED LION	
3 JOLLY WATERMAN	18 GREEN DRAGON	33 THE PHOENIX	48 RODNEYS ENGAGEMENT	
4 COOPERAGE	19 BLACK LION	34 THREE TUNS	49 BUNKERS HILL	
5 BLACK DOG	20 STAR INN	35 THE UNION	50 BULWARK	
6 DOLPHIN TAP	21 QUEENS HEAD	36 BRIDGE TAVERN	51 KINGS HEAD	
7 SEAHORSE	22 THE DRUMMER	37 HARE & HOUNDS	52 NEWCASLE-ON-TYNE	
8 WHITE LION	23 WATERMANS ARMS	38 TWO SAWYERS	53 QUEENS ARMS	
9 CROWN INN	24 RED LION	39 CROSS KEYS	54 OLD CRISPIN	
10 THE FOX	25 NORTHUMBERLAND	40 NAVY TAVERN	55 CROSS KEYS	
11 BLACK BEAR	26 THE CASTLE	41 DOVER CASTLE	56 FLOATING BRIDGE	
12 THE VICTORIA	27 MEDIATOR & PRIZES	42 BLACKSMITHS ARMS	57 THATCHED HOUSE	
13 THE BELL	28 HIGHLANDER	43 YORKSHIRE GREY	58 GEORGE & DRAGON	
14 STAR & GARTER	29 INDIA ARMS	44 KING OF PRUSSIA	59 PRINCESS ROYAL	
15 RODNEYS HEAD	30 ROEBUCK	45 ANGEL TAVERN	60 ARTICHOKE	

ALE-HOUSES
GOSPORT 1850

I wonder if this had anything to do with the decision to base the fledgling Submarine Service in Gosport?

Two
1904 to 1918

The Fort was expanded to house the growing Submarine Service and it became an independent command on 31 August 1912.

Various improvements were made in 1907 and 1913. The original Officers quarters, on the ground floor of Thames Block, were extended along the North Bastion. Officers were accommodated outside in wooden huts, now Lucia Block. The officer's heads were on the site of the old 'attack teacher'. In 1913 the more senior officers were moved from the North Bastion into the newly constructed Hazard Block, the 'empty caves' in the North Bastion, were taken over by junior officers.

Bonaventure Block was extended in 1917. ERAs Seamen and Stokers were housed in the ground floor of Arrogant Block. Petty Officers and Stoker Petty Officers were accommodated in the casement under the South Battery. Further additions were made to Arrogant block in 1918.

Onyx and Practolus Blocks were built in 1918. In 1908 Messrs Lipton opened a 'Temperance Restaurant' near the Rosario club.

Submarines from Fort Blockhouse were kept operationally busy during the war, while their home base hosted the country's first dedicated anti-submarine experimental station. The Nash fish-hydrophones were tested here, as was the first depth-charge thrower. A more novel approach to submarine warfare was taken with the introduction of Queenie and Billikins, two highly-trained sealions, who, it was hoped, would detect and then circle enemy U-Boats, uttering loud cries until friendly forces arrived to despatch the unwary foe. This praiseworthy idea came to an untimely end when Billikins went missing and was found in Stokes Bay, trying to gatecrash a mixed bathing party. So began a proud tradition that submariners have attempted to imitate over the years.

1917 was the year of the 'great petrol fire', vividly portrayed in W.L. Wyllie's painting which is hung in the fort. A number of C boats had been refuelling when somebody threw a lighted match over the side. Dockyard tugs were called in to help 'paddle' the water and fire away from the submarines and jetty.

New entry submariners undergoing basic instruction in 1904.

An early form of 'familygram'? 1904.

New entry submariners learning the ins and outs of the torpedo in 1904.

Instruction on the submarine battery in 1904.

D1 at Fort Blockhouse. Holland 3 is coming alongside.

The crew of a C Class submarine rehearsing for an early episode of 'Doctor Who' in their Rees-Hall escape suits.

Holland 5 aground off Fort Blockhouse 24 August 1910. The point where she ran aground claimed several unwary submarines over the years and, because of its highly visible position, became known as 'Promotion Point'.

Three

1918 to 1945

In 1930 the fort's discipline and administration were brought under one roof at the Fort Office Block, Alecto Colonnade. Some years later a CPO, accompanied by his trusty PO, walked into the 'Reg Office'. There were two leading regulators in the office, 'Okay' said the Chief, 'which of you is reading, and which is writing?'. This alleged incident took place in the late afternoon. 'I think, Chief, that you should go away for a while' boomed the Master At Arms from his office, somewhere in the depths of the Reg Office. The Chief and his PO left, only to reappear a short time later. 'All right,' said the Chief, 'You've had ten minutes, now which way round is it?!'

The Memorial Chapel, originally built in 1917, was greatly improved in 1929. In 1935 the canteen was converted to become the Church of St Ambrose.

Early in the Second World War, due to the unwelcome attention of the Luftwaffe, it was necessary to move the teaching role to Blyth. With the resourcefulness which only submariners can display, the fort's defences were improved. In more normal times the fort boasted two twelve-pounder guns on the ramparts which were manned by the Royal Artillery. At the start of the war a Bofors AA was added, later supplemented by two Low Angle six-pounders. This was a not a particularly impressive arsenal for the submariner to use to defend his home, the 4in submarine gun that was used for instructional purposes was pressed into service and mounted on the ramparts. More worryingly, the Engineering Department converted and manned two saluting guns. These fired a mixture of old iron and nuts and bolts, and were a well-intentioned early attempt at in-flight repair. A pom-pom, salvaged from a scrap heap in the dockyard, was mounted near the sub signal station, and was accompanied by a French machine gun on top of the signal station. Light guns were mounted on top of Forth and several other blocks; these were manned by Royal Marines. This fearsome armament was supplemented by an ample supply of rifles that were stowed around the fort, twenty-five Home Guard pikes and the 'Jones-Wise anti-tank gun'. This alarming weapon was fired by hitting a .22 cartridge, stuck in the vent of a section of metal drainpipe, with a hammer. Later, the lawn in front of the Wardroom was used to tether a barrage balloon, nicknamed 'Dolly Dolphin'. A searchlight manned by WRNS was positioned outside Arrogant Block. Perhaps because of this ad hoc, but nevertheless innovative, air defence system, Fort Blockhouse did not suffer as much damage as nearby Gosport and Portsmouth. However, it did not escape entirely. One night in January 1941, between 150 and 200 incendiary bombs fell on the fort, causing several fires. Luckily, or maybe or maybe as a result of the submariners AA barrage, the HE bombs that followed were all deposited into the surrounding water. These near misses caused two fatalities and several personnel received minor injuries. During the same raid, which was carried out by 300 enemy aircraft, 171 people were killed in Portsmouth, 430 were injured and 3000 left homeless.

Extensive mining preparations were also made. It was said that there were two tons of TNT under the roadway near the main gate. All the archways within the fort received similar treatment.

Admiral (Submarines) and his staff first moved to Fort Blockhouse after the First World War. They were housed in the CMB Base (now HMS *Hornet*). When HMS *Pandora* relieved HMS *Dolphin*, the Admiral and his staff moved on board, transferring later to a new office block in the fort. In 1939 RA (S) moved north to Scotland, to Aberdour near Rosyth, but moved again in April 1940 to a block of flats in north-west London called Northways. After the war RA(S) returned to *Dolphin*.

The Submarine school was a collection of dun-coloured huts situated where the 'tank' now stands. During 1941, due to the increased intakes of trainees, the school was moved north to a borstal institution in Blyth, which was re-named HMS Elfin. After the war the submarine training returned to HMS Dolphin.

During 1945 part of the moat outside the sea-front battery was converted into an open-air swimming pool. Rear Admiral G.E. Creasey (Admiral, Submarines) formally opened the pool in July 1946

Trainees undergoing escape training in the 'old tank', which was seaward of the present tank, during the early 1940s.

The 4in training gun at Fort Blockhouse.

Billiards and Snooker Room c1920.

Fort Blockhouse 1925. The moat can be seen clearly.

Fort Blockhouse main gate (Blockhouse Arch).

Instructional Staff at Fort Blockhouse 1926-1927.

King George VI at Fort Blockhouse December 1941, being greeted by Captain R.B. Drake.

A view of HMS *Dolphin* taken from the Brown area (Polish Camp).

Fort Blockhouse, photographed from HMS *Olympus* before her departure to Hong Kong in 1931.

WRNS at HMS *Dolphin* 1941.

WRNs in the Torpedo Workshop.

The South Bastion Wall.

Dredging the creek before the Second World War.

Aerial view of HMS *Dolphin*. The 'Brown Area' can be seen to the right of the picture.

Instructional Staff.

Part of the Torpedo Workshop.

Four

1945 to 1960

After the war submarine training returned to HMS *Dolphin*. Due to war time flotillas being paid off, manpower requirements were low. For the next twenty years the service was effectively 'all volunteer', a state of affairs which had not existed since 1925. Strangely, from the 1950s until the late 1970s the volunteer rate remained fairly steady at around 450. This number could support the conventional flotillas, but with the coming of the nuclear submarines intakes of 1200 a year were required. To meet this requirement 'direct drafting' was reintroduced. To deal with this large increase in trainees, accommodation at HMS *Dolphin* was expanded and plans were laid for a 'new school'.

I contacted several 'Dolphin old boys' during the preparation of this book, I think their recollections tell the story better than I could ever hope to:

'Blockhouse, or HMS Dolphin, was something to behold back in 1945 when I joined Submarines: grossly overcrowded, sleeping in hammocks wherever you could, canteen messing (cup of tea and a woodbine for breakfast). 103 Mess was affectionately called 'The Sheds'. 'The Stables' was a place beyond description – the electrics were unbelievable with bits of flex running here, there and everywhere for lighting and somewhere to plug in your iron. There were chalk marks on the deck, with a line to underneath your hammock so you could be shaken for the 'middle or morning watch' as Canteen or Area Patrol when on Duty Watch. Canteen messing, as I mentioned before, meant that as caterer of the mess I learnt how to make a 'clacker' and 'pot mess', jostling with the other caterers for space on the range for my mess's spuds and cabbage etc…them were the days!'

'We were Fort Additional in between submarines, working part of the ship, hoping to get picked for coaling party on a Friday so as not to have to attend Divisions, or dreading being picked for sullage party (pushing a cart around Blockhouse to the various galleys etc, eventually winding up at the sullage compound adjacent to Vulcan Block). We had long week-ends until 'first train Monday' and a railway warrant if we wanted one.'

'I came home from the Mediterranean and off foreign service leave as a killick to find my B13 for PO waiting in Blockhouse. I got rated and moved out of the 'Sheds' to the PO's mess alongside the parade ground and had my Burberry stolen the first night in the mess. I realized that as a killick you could only drink 'two and one' in case you got p****d, but the next day as a PO you can have "neaters".'

'I think I could ramble on for hours about dear old Dolphin from 1945 till 1969 especially as a young AB and LSEA. I also found Submarine Training very confusing, we trained in the classroom on 'S' boats went to sea during training on 'U' and 'V' boats and my first S/M was a 'T' boat!'

'I spent a lot of time in Dolphin off and on. I was there in 1959 for S/M course, then ran from there on Tireless before being spare crew. Then various courses, including the first S/M TASI course, onto Valiant for a while then back to Dolphin to instruct sonar courses. I'm afraid that photos weren't important at those times – the only one I have is a course photo of one of my officer classes. One event sticks in my mind; in spare crew we had the Blockhouse accommodation to look after (I was an AB). I was sweeper in the mess and my mate was church sweeper. We found an excellent place for soaking our socks – in the church font! The Padre caught us out and we got the sack; we were made sentries at Dolphin 2 as punishment. In '57 I was on the Adamant in Rothesay Bay, then shifted round to Faslane. I worked as an operator on the ship's telephone exchange; the shore-side exchange was a wood hut on the jetty manned by three civvies. In those years the Gareloch was full of mothballed ships: Rodney, Renown etc. I was acquainted with a civvy who had the contract to clear the decks of bird droppings.

The Submarine Escape Training Tank (SETT) was opened in 1954 and the first training class put through on 13 July 1954.

HMS Dolphin in the late 1950s.

This is the base plate that eventually grew into......

...this, which grew into......

...this, which eventually turned into......

...this. The Tank nears completion 16 April 1952.

The Tank as completed in 1953.

A well deserved break, by the bottom section of the 100ft tower 6 November 1950.

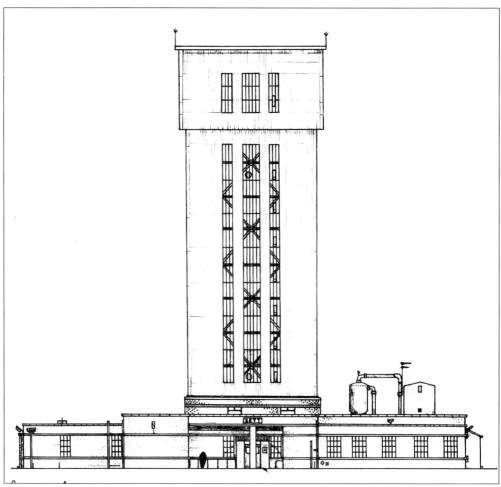

The tank was officially opened on 13.July 1954. It is thirty metres deep and hold 20,000 gallons of water. Personnel from fifteen foreign and Commonwealth Navies have been trained over the years, with a peak average of 4,500 students a year in the 1960s and 1970s

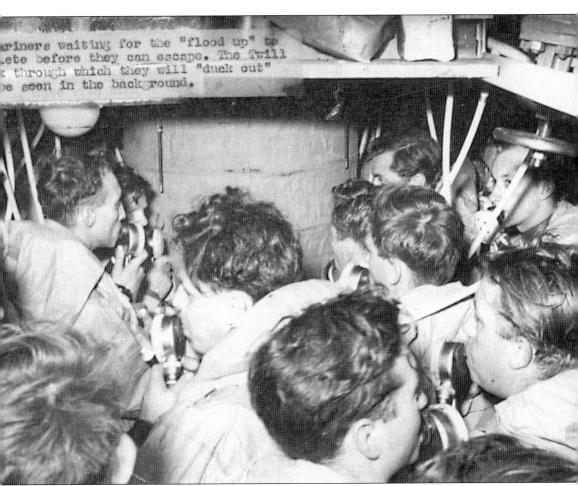

riners waiting for the "flood up" to
ete before they can escape. The Twill
through which they will "duck out"
e seen in the background.

Rush escape drills from the 100ft section 1954.

Senior Rates in the Chief and Petty Officers Mess, Hazard Block, Christmas 1946 or 1947.

The Dolphin Summer Ball 1950.

Winter at Fort Blockhouse Main Gate.

BonAdventure and Hazard Blocks.

An aerial view of Blockhouse taken between July 1945 and February 1946. HMS *Rorqual* is outboard with an X-craft just visible outboard of her. Another X-craft can be seen on the other side of the jetty.

Dolphin, aerial view, 1952.

Looking
east 1954.

Looking north from the South Bastion.

View from the Southern casement looking towards Portsmouth.

The Southern Casement.

Senior Rates Mess under construction June 1956.

The new Guardhouse – before the Guardhouse was there. June 1956.

Accommodation Blocks under construction June 1956.

Busy construction scene July 1956.

Accommodation blocks from the Tank.

No 1 Block nearing completion – September 1956.

Senior Rates Mess September 1956. The picture appears to have been taken after a particularly good Mess Dinner.

HMS *Dolphin* Main Gate February 1957, the gate house is there now.

Aerial view of the jetty prior to the start of the jetty reconstruction September 1959.

Pier Head and the north-east end of the Jetty. The moorings for the SPUD pontoon are being laid.

Looking towards the Main Gate from the balcony of the Senior Rates Mess.

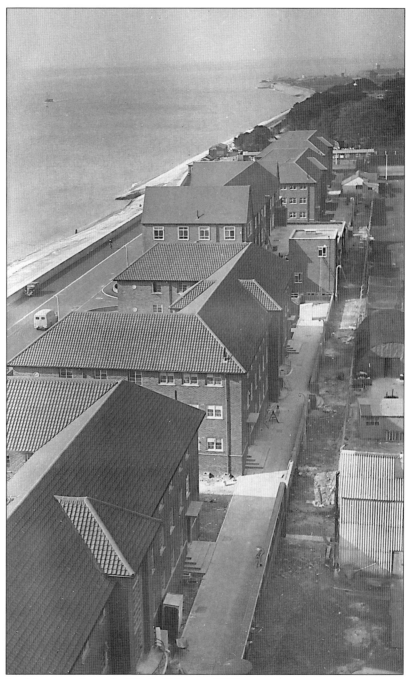
Looking west from the Tank, February 1957.

North-east end of the jetty.

Pile driving on the jetty approach. The building on the left is built on top of the original fort wall.

Jetty approach under construction.

Laying prestressed pre-cast beams and filler blocks on the approach the tank can be seen in the background, September 1959.

Placing main beams at north-east end of jetty October 1959.

Placing main beams at north-east end of jetty October 1959.

Decking the jetty approach October 1959.

Decking the jetty approach October 1959.

Jetty approach October 1959

General view
of north-east
section.

Western end of Jetty
approach October 1959.

Looking north-east December 1959.

Looking north-east December 1959.

FOSM Staff (not dated).

FOSM Staff 1955.

Aerial view taken in the early 1950s.

The Floating Dock 1946.

Five

1960 to 1998

Further improvements were made to the accommodation at HMS *Dolphin* during the 1970s; the new Junior Rates accommodation and messes were built around the playing fields.

The various reviews of both naval manpower and naval establishments led to a reappraisal of the role of HMS *Dolphin*. The base was decommissioned on 30 September 1998 and reopened as the Defence Medical College on 1 October.

It is hard to predict what the future may hold for the fort. The Defence Medical College is now to be moved, and there seems little chance that the events of 1904 will be repeated.

So ends the story of HMS *Dolphin* – I've told it as best I can. I hope that it has brought back many a pleasant memory and as someone once said, 'You can check out any time you like but you can never leave' . . .

The in-built plaque to
mark the completion
of the jetty
reconstruction.

March 1960 Completion
ceremony with Director
General in attendance.

Then after all that hard work ..... at half past three on 14 April 1960 HMS *Auroces* collided with the Jetty.

Damage to the handrails and jetty edging. Another victim for Promotion Corner? I've heard tell that Motor Room watch-keepers would try and guess who was bringing the submarine alongside by the number of 'bumps'. I wonder if anybody got this one right?

Dolphin infill, boring in the trench bottom ahead of a ramp to receive plenometer 14 June 1966.

Dolphin infill start of the ramped embankment.

General view of ramp 14 July 1966.

Foreshore protection 27 September 1966.

General view from the shore end of the Petrol Pier.

General view looking south-west 6 April 1967.

General view looking East 25 August 1967.

Foreshore Protection at the Eastern end of the filled area 19 August 1966.

12 May 1967.

General view of TAS showing the Spud.

View looking towards Pandora Block prior to the start of the coastal protection work 3 November 1965.

Foreshore protection slip No 5 at the junction of the spur and ramp 28 June 1966.

General view 3 November 1965.

28 June 1966.

Looking towards Pandora Block 3 November 1965.

General view of the Spud from the existing jetty 12 May 1965.

TAS jetty 12 May 1965

TAS Jetty. General view of the site after completion of laying the concrete deck unit. 21 Sept 1965

Foreshore. 16 September 1974.

From the foreshore looking towards the Tank. 16 September 1974.

Looking towards the Petrol Pier. 19 August 1966.

General site view. 21 July 1977.

General view, 2 December 1977 Haslar Bridge (Pneumonia Bridge) can be seen in the background.

Infill 15 June 1967.

Infill 14 July 1966.

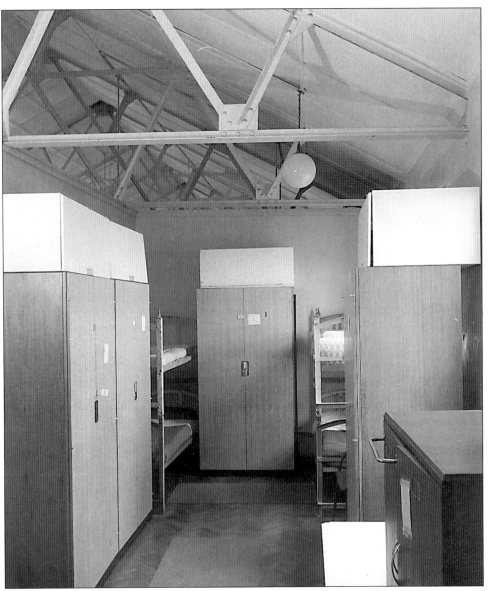

1970s Junior Rates Accommodation. Top Floor Arrogant Block.

Arrogant Block Top floor central passageway.

Pactolus Block Junior Rates accommodation. The windows were shuttered against damage from flying shingle and sea spray in heavy weather.

Arrogant Block on the left perilously close to the Explosive Storehouse in the South Casements. Alecto Block, Rapid Block and No 1 Sub station are on the right.

PSTO(N) lorries unloading in the North Bastion.

Modern bathrooms built against the original East wall of Arrogant Block. The North Bastion is just visible in the background.

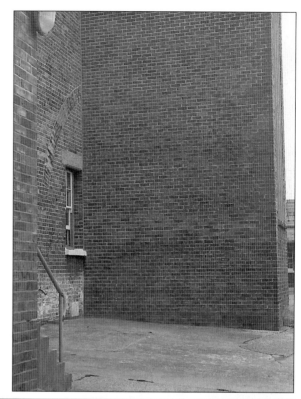

Below: Looking north between Clyde Block and the North Bastion 1970s.

Above: Entrance to the North Bastion, Thames Block to the left Clyde Block is on the right.

Below: The four archway supports to the firing gallery are partially obscured by the Clyde Block entrance porch.

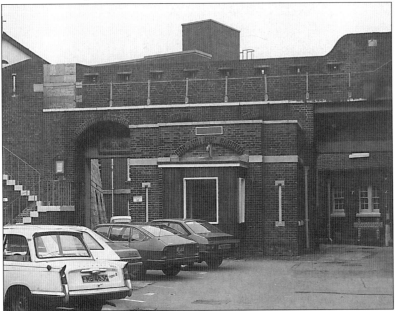

The two photographs on the opposite page show the installation of the single tower during 1965, pictured through the roof of the tower.

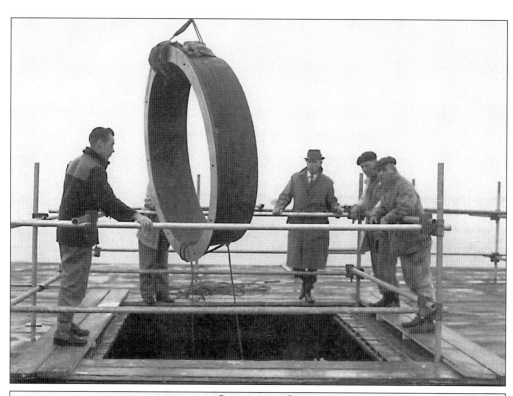

Above: Lowering the Single tower into the 100ft-tank section.

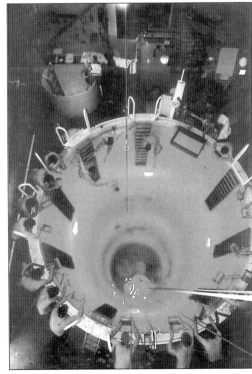

The Tank top, looking down

95

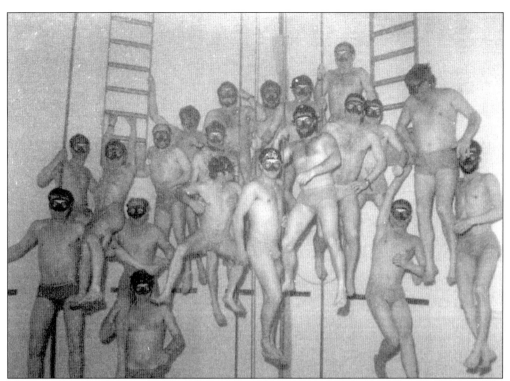

The tank staff being.....well, the tank staff.

Floor plan of the tank. Despite their rugged exteriors, the tank staff always thought the building was too cold.

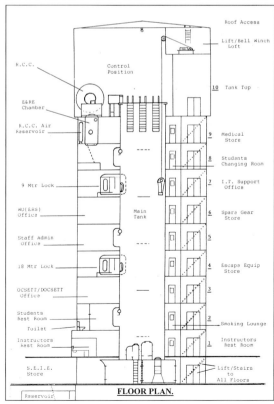

Roof Access

Lift/Bell Winch Loft

R.C.C.

Control Position

E&RE Chamber

R.C.C. Air Reservoir

9 Mtr Lock

WO(ERS) Office

Staff Admin Office

18 Mtr Lock

OCSETT/DOCSETT Office

Students Rest Room

Toilet

Instructors Rest Room

S.E.I.E. Store

Reservoir

Main Tank

10 Tank Top

9 Medical Store

8 Students Changing Room

7 I.T. Support Office

6 Spare Gear Store

5

4 Escape Equip Store

3

2 Smoking Lounge

1 Instructors Rest Room

Lift/Stairs to All Floors

FLOOR PLAN.

Eventually the tower was insulated to give it the appearance it has today.... bless 'em.

The pictures on these pages were photographed in June 1978

Thames Block frontage.

The North Bastion is in the centre of the photograph with Thames Block to the left and Arrogant block to the right. Clyde Block is behind the bastion. 1970s.

The west end of Thames Block (right).

The South Casement as seen from the approaches to Portsmouth Harbour.

HMS *Dolphin* frontage, as seen from across the harbour entrance at Portsmouth Camber.

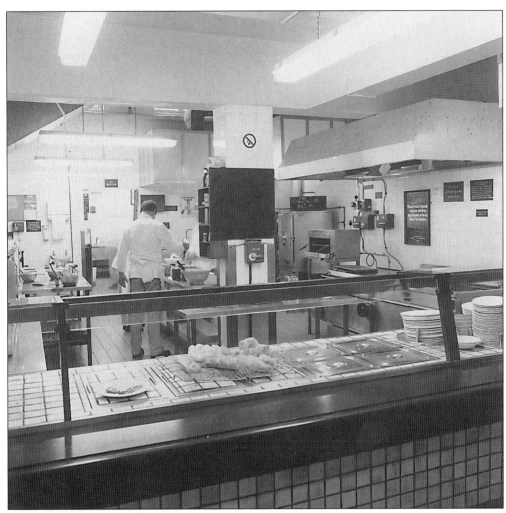

Main Galley Hazard Block.

All mod cons? Junior Rates bathroom.

Signal Tower on the South Casement.

The following five pictures were photographed in June 1978

Nov 1978.

The road to Haslar Jetty, 'Dolphin 2' is on the left. May 1975.

May 1975.

Haslar Jetty, this site is now occupied by the Submarine Museum.

General site view

The following pictures were photographed in September 1985.

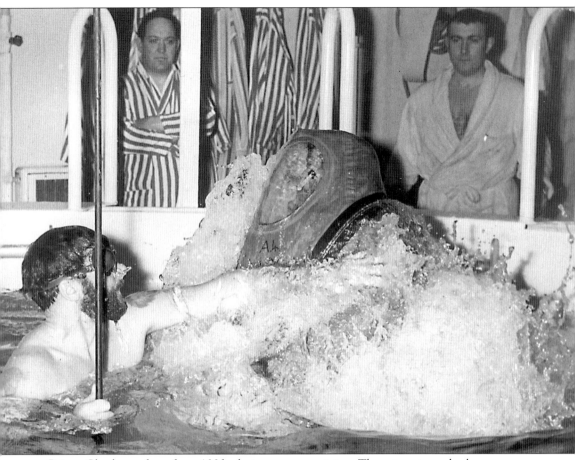

Prince Charles surfaces from 100ft, during escape training. There is not much else one can say about this picture other than to refer readers to the anecdote in the introduction.

Six

Submarine Squadrons

On arrival at Portsmouth in 1901, the four Holland class submarines and their attendant depot ship, HMS *Hazard*, were ordered, because of their dangerous petrol engines, to station themselves in the far reaches of the harbour, a paranoia that will be familiar to the modern-day nuclear submariner. The fledgling squadron moved to Fort Blockhouse in 1904, taking it over from the Royal Engineers.

When Commodore Roger Keyes took command of the submarine flotilla in 1911, his main aim was to weld the few, mainly experimental. craft, into an effective fighting force. This was made easier due to the fact the First Sea Lord of the Admiralty, Sir John Fisher, thought that submarine warfare would revolutionize naval strategy and was intent upon acquiring a large submarine fleet. Other senior officers did not share his enthusiasm for a strong submarine force and, apart from sanctioning the increased building programme, the infant submarine flotillas received little help from the Admiralty. It should be remembered that at this time submariners were considered ungentlemanly, unwashed and prone to dress as 'North Sea fishermen'. Certain British Admirals were calling for them to be hung! This attitude, if unforgivable, is understandable. At this time there were no senior officers with any submarine experience. The Admirals, schooled in the Victorian discipline and class-ridden traditions of the Royal Navy, were unlikely to take much notice of the very young men and junior officers who were put in command of the crude and small submarines. It should be remembered that these same gentlemen said of the embryonic Fleet Air Arm: 'Their Lordships are of the opinion that they would not be any practical use to the Naval Service.' So they knew what they were talking about some of the time.

So, at the start of the First World War in 1914, it might not be an understatement to say that Britain's infant submarine service did not enjoy any great popularity at the Admiralty. Yet the Royal Navy, at that time the biggest navy in the world, had much to fear from the submarine. Quite apart from the need to protect the far-flung reaches of the Empire, Great Britain was dependent on the sea for almost all of its food and material needs. A single submarine of a few hundred tons could destroy many thousands of tons of merchant and naval shipping, while a determined enemy submarine force could be capable of bringing the might of the British Empire to its knees, as nearly happened in both World Wars.

It was left to the submarine commanders themselves – men such as Max Horton, Martin Nasmith and Noel Laurence – to develop the tactics of submarine warfare, and dispel once and for all the myth that submarines should only be used for coastal defence. Despite the difficulties of mastering unfamiliar and often dangerous craft and, as mentioned above, gaining the acceptance of a disapproving audience, these men, through personal courage and imagination, managed to develop a model fighting force.

At the outbreak of the First World War the HMS *Dolphin* flotilla, then the 2nd Flotilla commanded by Captain Roger J.B. Keyes, consisted of the following submarines:
E10 – Lt-Cdr William St J. Fraser
D2 – Lt-Cdr Arthur G. Jameson
S1 – Lt-Cdr Gilbert H. Kellett
A5 – Lt. John de B. Jessop
A6 – Lt. Henry E. Smyth
A13 – Lt Edward R. Lewes
B1 – Lt Basil A. Beal

Britain had around fifty-five serviceable submarines at the outbreak of the Second World War, including twelve boats (H and L class) used for training. These dated from the 1914-1918 period. Some of these – the H class boats – were used operationally in 1940 but withdrawn soon afterwards, following two early losses. An emergency building programme was instigated and 168 were subsequently built. Seventy-eight were lost during the war.

Submarines attached to HMS *Dolphin*, then known as the 5th Submarine Flotilla, carried out patrols in the Channel and Bay of Biscay during the war. X-craft from HMS *Dolphin* marked the beaches for the Normandy landings.

A Holland class submarine alongside HMS *Hazard*.

H. M. S. "Thames" Mother of the Submarines

HMS *Thames* lying in the upper reaches of Portsmouth Harbour, 1904.

The first submarine coxswain in the Royal Navy, Petty Officer William Waller. In common with most other early submariners, he was a volunteer and a very senior one at that.

Captain Reginald Bacon RN, who in 1905 became the first inspecting Captain of Submarines and as such was the head of the fledgling submarine service.

Holland 3 passing HMS *Victory* in Portsmouth Harbour.

The next four pictures show busy jetty scenes

HMS *Oberon*, one of the highly successful Oberon Class.

A member of D.2's ship company demonstrating that certain indefinable pride that submariners have in their appearance.

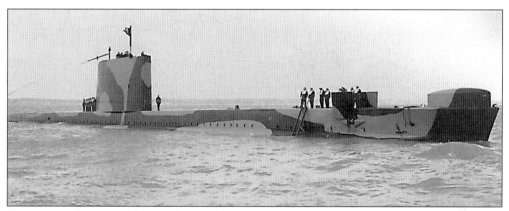

HMS *Otus* entering Haslar Creek, flying her paying off pennant. Her unusual paintwork, duck egg blue and black, a colour scheme adopted by British submarines in the Far East during the Second World War, was applied for her time during the Gulf War – hence her Jolly Roger.

On a fine summer's evening, 1 July 1971, HMS *Artemis* sank in 30ft of water alongside at HMS *Dolphin*. Three of the crew were trapped onboard for ten hours but finally escaped in the early morning.

Senior Offices from the Baghdad Pact boarding HMS *Alcide* during a visit to HMS *Dolphin* August 1959.

HMS *Amphion* alongside at HMS *Dolphin* 1969.

HMS *Spartan*, the last submarine to salute HMS *Dolphin*, days before it was decommissioned.

First the Nab. And then the Warner. Outer Spit. And Blockhouse Corner.